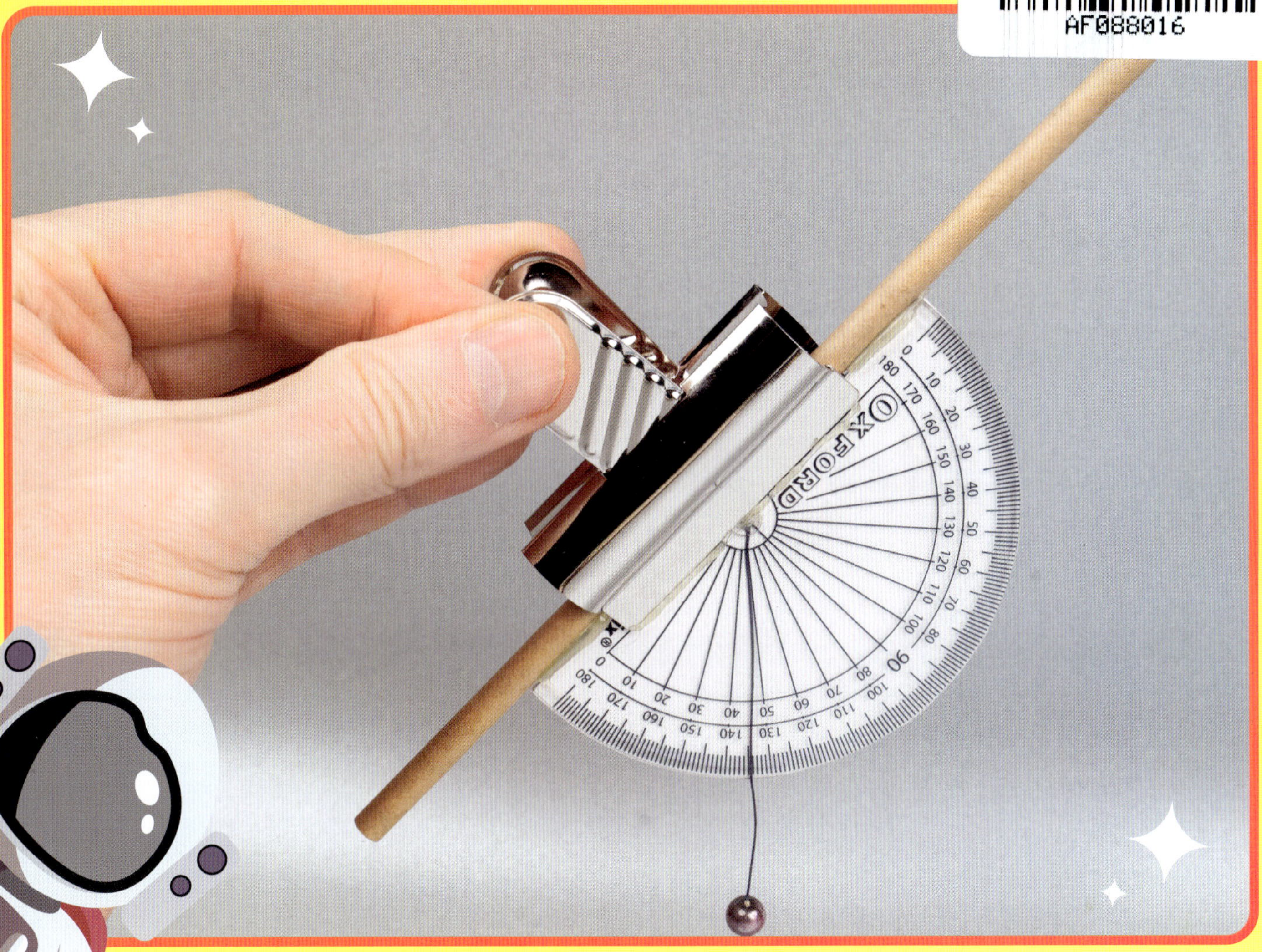

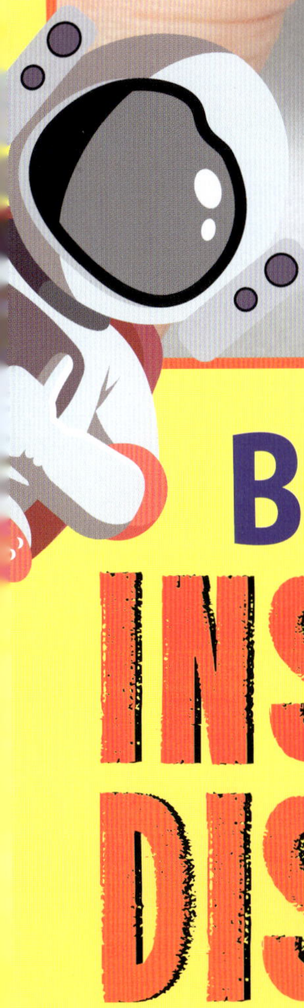

BUILD YOUR OWN
INSTRUMENTS TO DISCOVER SPACE

Rob Ives

Words in **bold** can be found in the glossary

CONTENTS

PREPARE FOR LIFTOFF	4	SAFETY FIRST!	6

INSTRUMENTS TO DISCOVER SPACE	8	NOCTURNAL TIMEPIECE	20
INCLINOMETER	10	CONSTELLATIONS	22
MOON PHASES	12	EQUATORIAL SUNDIAL	24
REFLECTING TELESCOPE	15	SUPER DISTANCES	28
REFRACTING TELESCOPE	18	THE HUBBLE TELESCOPE	30

GLOSSARY AND INDEX	32

Copyright © 2024 Hungry Tomato Ltd

First published in 2024 by Hungry Tomato Ltd
F15, Old Bakery Studios, Blewetts Wharf, Malpas Road, Truro, Cornwall,
TR1 1QH, UK.

No part of this publication may be reproduced, stored in a retrieval system, or transmitted in any form or by any means, electronic, mechanical, photocopying, recording, or otherwise, without prior written permission of the copyright owner.

A CIP catalogue record for this book is available from the British Library.

ISBN 9781916598874

Printed in China

Discover more at
www.hungrytomato.com

PREPARE FOR LIFTOFF

Try your hand at building amazing space-themed models! Using smart and simple engineering principles, you can make a whole collection of out-of-this-world crafts that show the wonders of our universe and beyond!

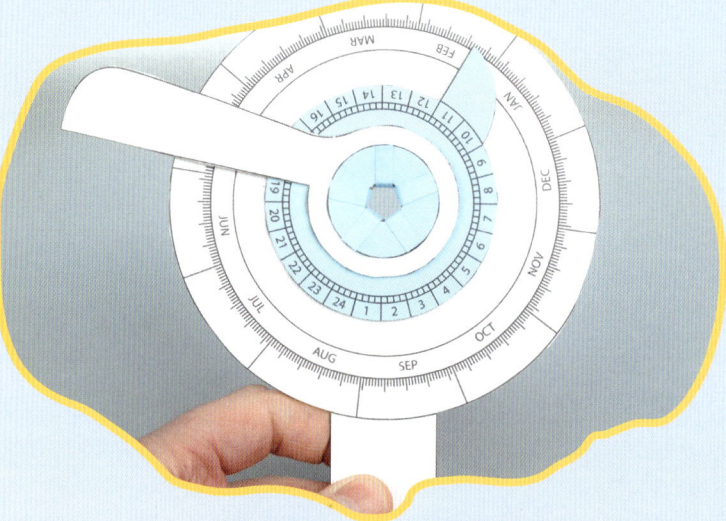

THIS BOOK IS INTERACTIVE!

Some of the projects in this book come with templates to help you cut pieces to the right shape and size. Use a smartphone to scan the QR code at the beginning of the project to access a downloadable template that you can print out.

You will find QR codes at the end of some projects, too. These will direct you to videos of the moving models in action!

You can also find all templates and videos at:
www.hungrytomato.com/instruments-to-discover-and-study-space

PREPARE FOR LIFTOFF

TOP TIPS

- Before you start any project, read the step-by-steps all the way through to get an idea of what you are aiming for. The pictures show what the steps tell you.

- When printing templates, check that your printer is set to "print to scale" or to "full size" to make sure they come out the right size for your other materials!

- Use a cutting mat, or similar surface, for cutting lengths of craft sticks, skewers, and anything else you may need.

- Use the sharp end of a pencil to make small holes in cardboard (see page 7 for method) or ask an adult to help, using either scissors or a craft knife.

- Where strong glue is required, you may want to use a glue gun. Make sure you ask permission, and do not use it without an adult present. Strong liquid glue, such as wood or epoxy glue, will work well, too.

★ EASY

★★ MEDIUM

★★★ HARDER

You will find stars in the corner of the first page of each craft. These stars are a guide to the difficulty level of each project. They show you when you may need another pair of hands!

SAFETY FIRST!

Be careful and use good sense when making these models. They are easy to understand but will require some cutting, gluing, drilling, and other awkward tasks that you may need some help with from an adult.

WHEN TO GET HELP

Watch out for this sign throughout the book. You may need help from an adult when completing these tasks.

DISCLAIMER

The author, publisher, and bookseller cannot take responsibility for your safety. When you make and try out the projects, you do so at your own risk. Look out for the safety warning symbol (shown above) given throughout the book and call on adult assistance when you are cutting materials or using a pair of scissors or pliers, craft drill, or hot glue.

SAFETY FIRST!

HOW TO CUT A POSTER TUBE SAFELY:

1.
Cut a strip of cardboard and fold a right angle into it. Measure from the crease the width you need the tube to be and make a hole at that point.

2.
Hold the cardboard over the end of the poster tube. With a pencil in the hole, twist the tube around to draw a line parallel to the edge.

3.
Ask an adult to carefully cut along the line to make a short section of tube. They could use scissors or a craft knife.

HOW TO MAKE HOLES IN CARDBOARD SAFELY AND EASILY:

Pressing a pencil point through cardboard and into an eraser, like the photo on the right, is a safe and easy way to make holes.

INSTRUMENTS TO DISCOVER SPACE

Humans have studied outer space for centuries. In fact, scientists have found cave drawings of the Sun and stars which date back tens of thousands of years!

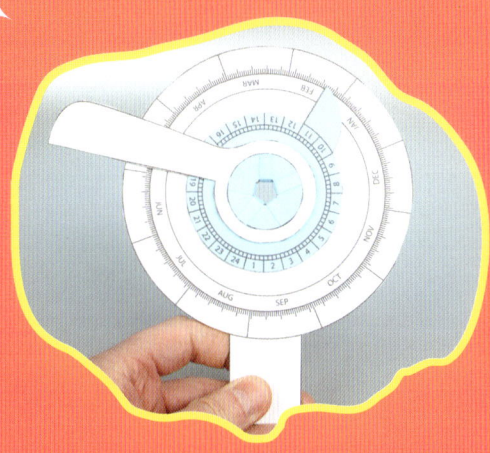

OBSERVING SPACE

The first **astronomers** – people who studied outer space – didn't have the technology we have today. They relied on their eyes to observe the movements of the Sun, Moon, and stars. They noticed patterns in these movements and found clever ways to record them.

TELESCOPE DISCOVERY!

For years, it was thought that the **Sun orbited** Earth. It wasn't until the 1600s, when telescopes were used to see faraway planets, like Jupiter, that we discovered all planets in the solar system orbit the Sun. This discovery was a real breakthrough!

WANT TO KNOW MORE?

This book is full of fantastic projects which will allow you to conduct your own studies of space. Turn the page to dive in - who knows what you'll discover?

INCLINOMETER

The inclinometer is a scientific instrument used to work out the height of stars in the sky by measuring the angle between the star and the horizon.

WHAT YOU NEED:
- Thick card
- Bulldog clip
- Paper straw 6mm
- Glass bead 6mm
- Thread/thin string
- Toothpick

TOOLS:
- Pencil
- Semi-circular (180 degree) protractor
- Pair of scissors
- Ruler
- Glue gun (cool melt) or strong craft glue
- Push pin or small craft drill

1 Trace the shape of your protractor onto thick card and cut out.

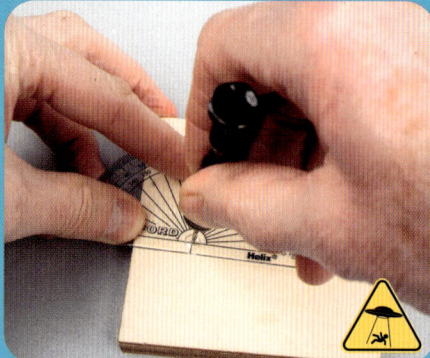

2 Ask an adult to make a fine hole in the protractor at the middle point, using a push pin or small craft drill.

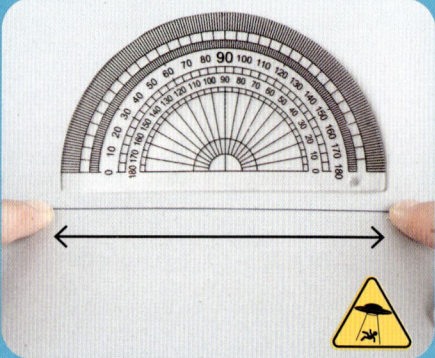

3 Cut a piece of thread to the length of the base of the protractor.

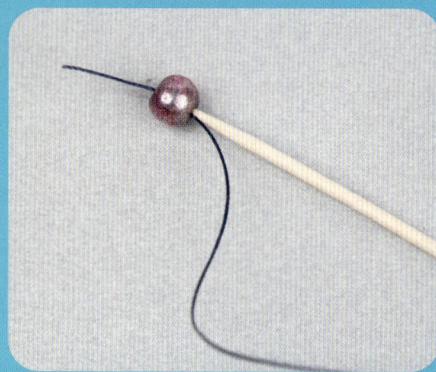

4 Push the thread through the bead. Secure it in place by pushing a toothpick into the hole.

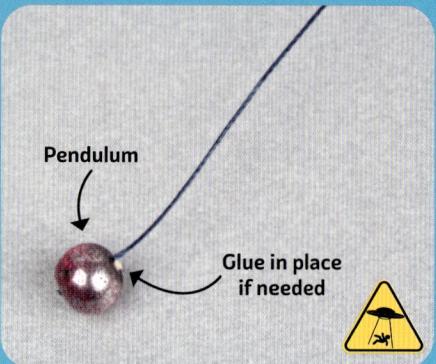

5 Using scissors, carefully cut off the excess toothpick, leaving the tip inside the bead as shown. This bead will be your pendulum.

6 With the bead hanging down the back of the protractor, push the thread through the hole.

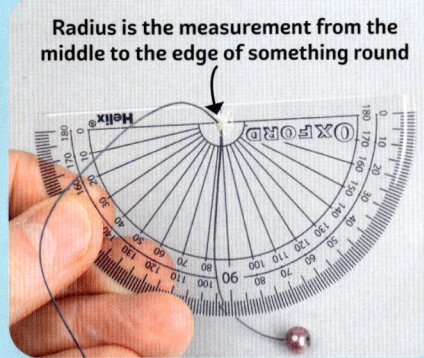

Radius is the measurement from the middle to the edge of something round

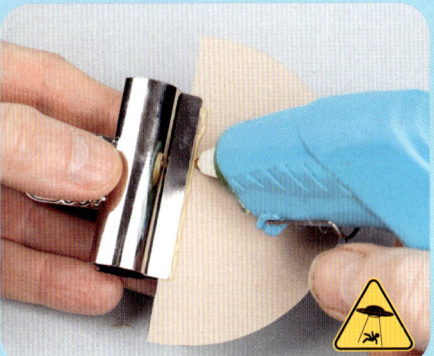

7 Measure the thread so it's 25mm longer than the protractor's radius. Then, glue over the hole to secure.

8 Cut off any excess thread. Then, grip the protractor with the bulldog clip, gluing the front jaw along the length of the protractor.

9 Fit the cardboard semi-circle in place at the back of the protractor. Glue it to the back bulldog jaw to sandwich the pendulum.

10 Thread the paper straw through the bulldog clip so that its middle roughly lines up with the middle of the protractor.

How to use your inclinometer:

1.
Squeeze the bulldog clip so that the pendulum hangs free.

2.
Sight the star you're measuring through the straw.

3.
Release the bulldog clip to trap the thread.

4.
Read the protractor angle that the pendulum is lined up with.

5.
Subtract this number from 90 to get the star's inclination.

DID YOU KNOW?
Inclinometers have been used for centuries: they're vital when using the stars to navigate, making maps of the sky, and recording star positions.

MOON PHASES

Watch the progression of the phases of the Moon in super speed with this mini model.

Use the QR code to access the template you need.

WHAT YOU NEED:
- Poster tube 80mm in diameter
- Table tennis ball
- Corrugated cardboard
- Assorted card
- Bamboo skewer

TOOLS:
- Pencil and eraser
- Black marker pen
- White marker pen
- Ruler
- Pair of scissors
- Strong craft glue
- Craft drill

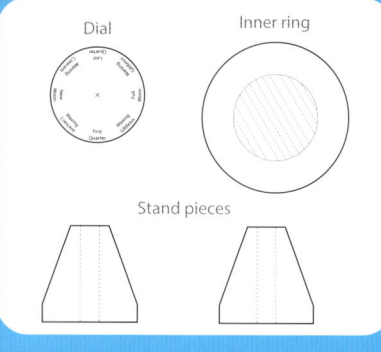

1 Print, copy, or trace the shapes from the template onto the specified materials and cut out.

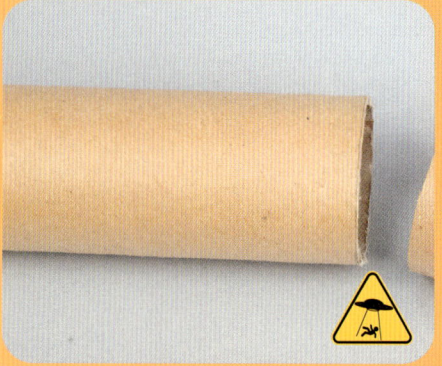

2 Ask an adult to cut a ring out of a poster tube (see page 7). It needs to be as wide as the table tennis ball.

TOP TIP
If you don't have a poster tube, you could use a wide masking tape roll!

3 Cut a strip of corrugated cardboard half the width of the tube and long enough to wrap around the interior completely. Glue it to the inside of the tube. It should be level with the top of the tube and reach halfway down.

4 Paint the inside of the tube without the card insert or use a black marker pen as shown.

5 Fit and glue the inner ring into place against the corrugated cardboard lip.

6 Ask an adult to drill a hole halfway down the tube, behind the inner ring. Make another directly opposite.

7 Ask an adult to drill a hole at each end of the table tennis ball. They should be on the seam.

8 Cover one half of the ball in black marker pen, using the seam as a guide.

9 Place the ball inside the tube and line up the holes. Then, thread the skewer through the tube and ball. The ball should be free to rotate. Trim the skewer so that 13mm remains at each side.

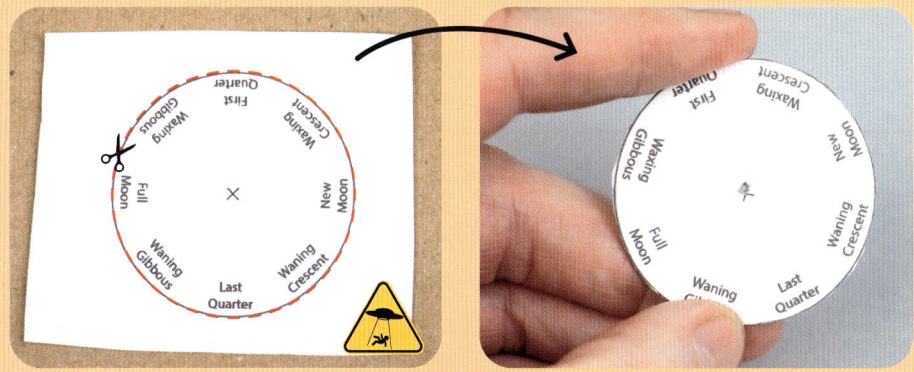

10 Glue the card dial from the template to a piece of corrugated cardboard and cut out around the circle. Then, make a hole through the cross in the middle (see page 7).

11 With the text facing up, slide the dial onto the skewer, under the poster tube as shown.

TOP TIP

The dial and ball should fit perfectly and twist with the skewer. If they don't, you can glue them onto the skewer. Be careful not to glue the poster tube though!

Halfway

12 Arrange the model so that the black half of the tube, white side of the ball and "full moon" are facing the front.

13 Glue the stand legs about halfway up on each side of the poster tube. Make sure that the skewer and dial are off the counter as shown.

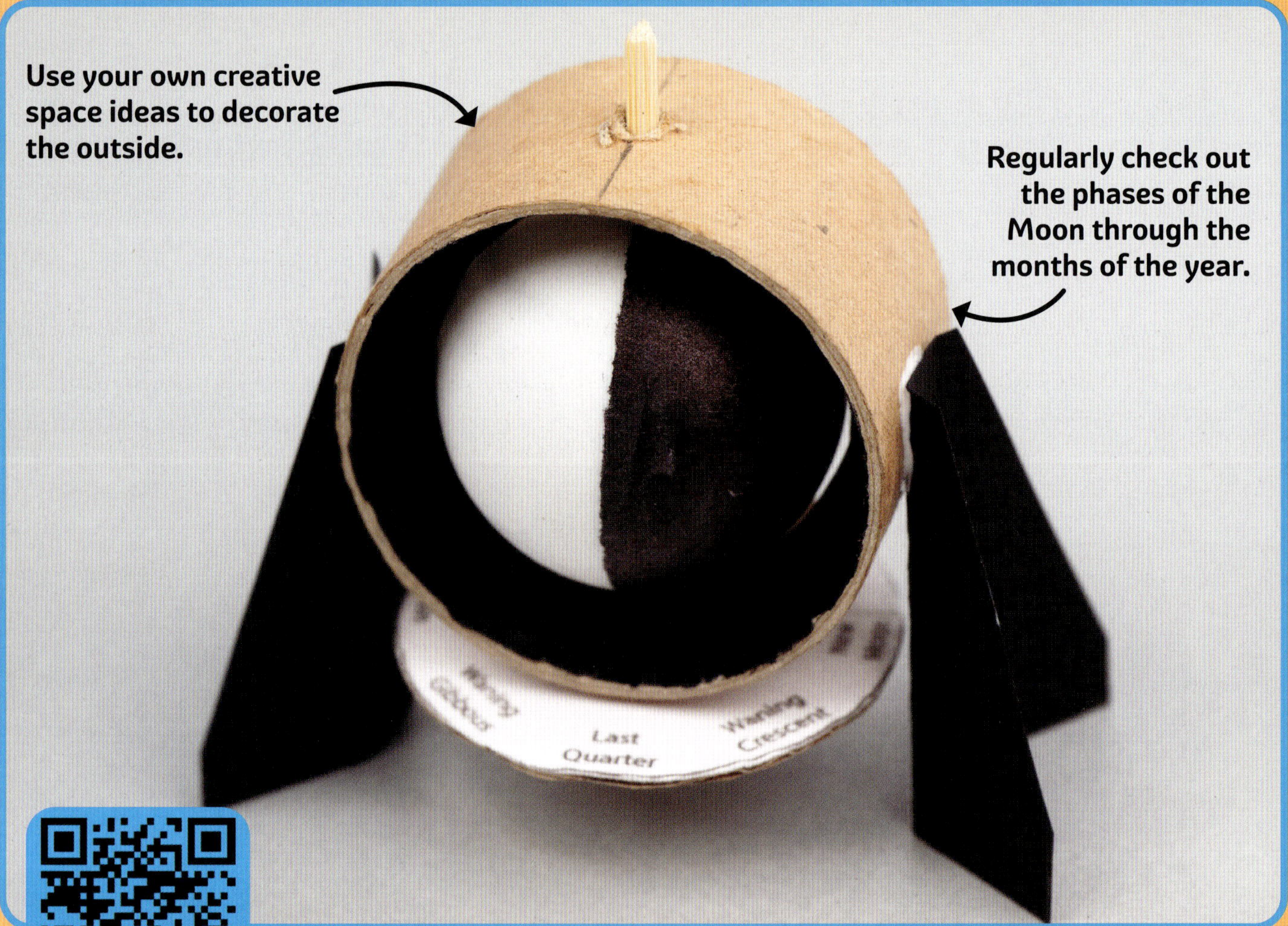

Use your own creative space ideas to decorate the outside.

Regularly check out the phases of the Moon through the months of the year.

ROTATE THE DIAL TO SEE THE DIFFERENT PHASES OF THE MOON!

REFLECTING TELESCOPE

Search for stars, moons, and undiscovered planets with this amazing reflecting telescope. What will you discover?

Use the QR code to access step-by-step instructions for step 2.

WHAT YOU NEED:
- Compact magnifying mirror
- Small, flat mirror
- Magnifying eye glass
- Corrugated cardboard
- Pipe cleaners
- Bamboo skewer
- Assorted card
- Small paper ice-cream tub
- Paper straws 200mm

TOOLS:
- Pair of scissors
- Strong craft glue
- Paper packing tape
- Compass
- Ruler

1 Using a pencil, trace the compact mirror onto corrugated cardboard and cut out. Repeat to make a second circle.

2 Mark one of the cardboard circles to show the thirds. **Need help with this? Scan the QR code above to find out how to mark thirds.**

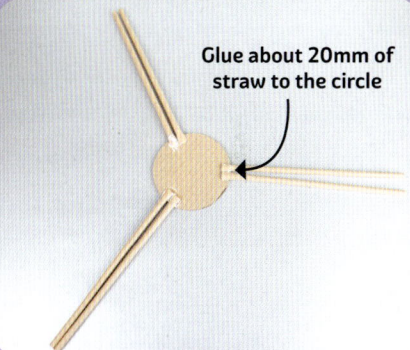

3 Glue pairs of paper straws on the marks as shown. Then, glue the second cardboard circle on top, covering the straw ends.

4 With the mirrored side facing up, glue the compact mirror to the top cardboard circle.

5 Connect each straw to the one next to it to create triangle shapes as shown. Insert a pipe cleaner into the straw ends to hold them together.

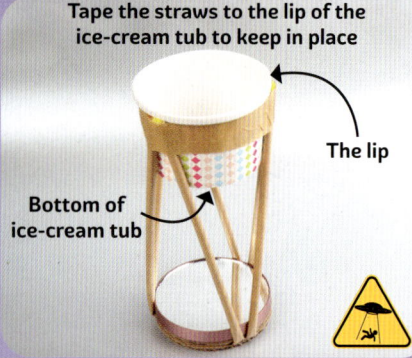

6 Ask an adult to cut the bottom out of the ice-cream tub and place in the middle of the straws and central over the mirror.

TOP TIP

These craft items are a little bit harder to find. Here's a few ideas of where to look:

- Magnifying compact mirrors can often be found in pharmacies and beauty stores.
- The big magnifying glass lens - used in the Refracting Telescope (page 18) - can often be found online or in charity shops.
- Small magnifying eye glasses can often be found online, or in craft shops and charity shops.

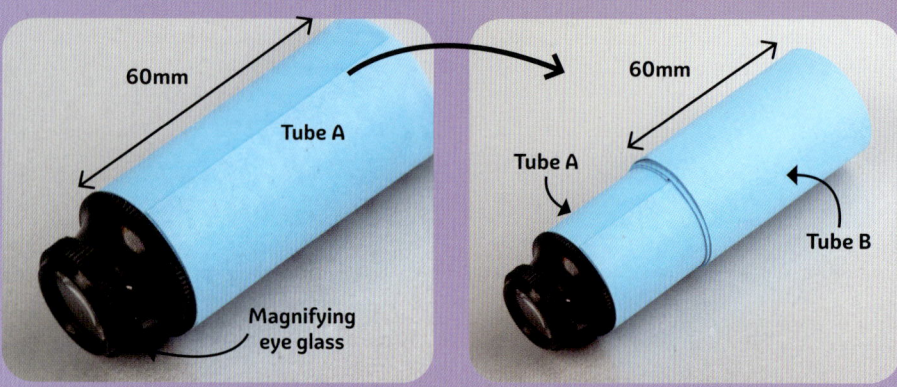

7 Wrap a card tube (tube A) tightly around the magnifying eye glass. Glue the loose end down as neatly as possible. Then, make a second tube (tube B) to fit over tube A, as shown, so that they can slide in and out of each other.

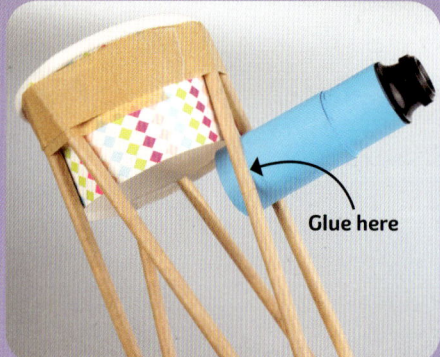

8 Glue tube B between two straws, just below the ice-cream tub as shown.

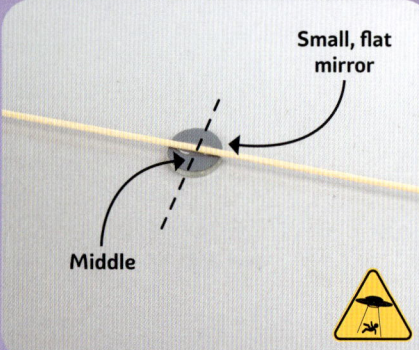

9 Cut a bamboo skewer to 140mm long. Glue the middle of the skewer to the back of the small, flat mirror.

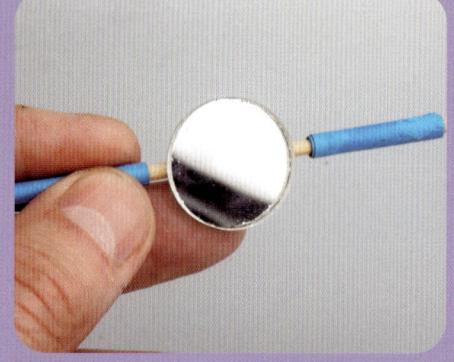

10 Wrap paper tightly around each end of the skewer and glue down the ends. Make sure the skewer can still twist.

11 Position the skewered mirror under the ice-cream tub, facing and in line with the magnifying eye glass, as shown.

How to use your reflecting telescope:

1. Look through the magnifying eye glass.

2. Adjust the small mirror to the right position so you can see.

3. Pull tube A back and forth to focus the image.

TOP TIP

Remember, it's NEVER safe to look at bright lights or the Sun, so don't point your telescope directly at them.

WHAT CAN YOU SEE?

LOOK THROUGH HERE!

1. Magnifying eye glass
3. Tube A
2. Small, flat mirror

DID YOU KNOW?
There are two main types of telescopes: refracting (see page 18) and reflecting, which is best for brighter things like moons and planets.

17

REFRACTING TELESCOPE

Refracting telescopes are great for seeing things from a long distance. What faraway things can you spot using your amazing refracting telescope?

WHAT YOU NEED:
- Paper cup
- Magnifying glass lens
- Magnifying eye glass
- Poster tube
- Corrugated cardboard
- Assorted card

TOOLS:
- Pair of scissors
- Strong craft glue
- Ruler

1 Ask an adult to cut the base out of a paper cup. Fit the magnifying glass lens into the wide end of the cup. Glue in place if it doesn't fit tightly.

2 Roll up a tube of card so that the magnifying eye glass fits neatly into it. Glue the loose end down as neatly as possible.

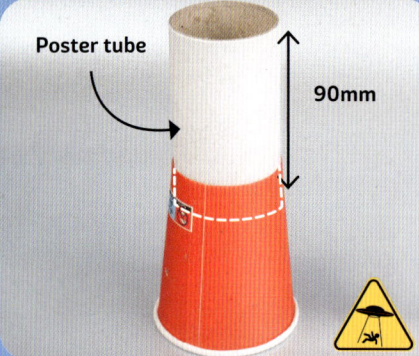

3 Ask an adult to cut the poster tube. Slide it into the small end of the paper cup until it's snug. Use glue to secure together.

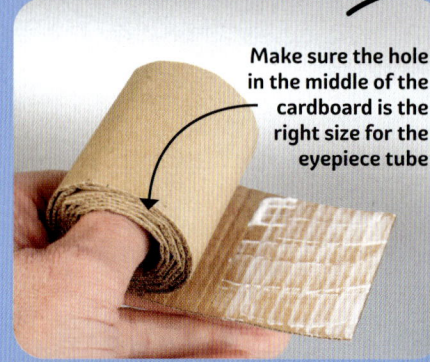

4 Cut a length of corrugated cardboard that's the same height as the poster tube. Roll up the cardboard, leaving a hole in the middle for the eyepiece, and glue down the end to stop it unwinding. Fit the rolled up cardboard into the poster tube.

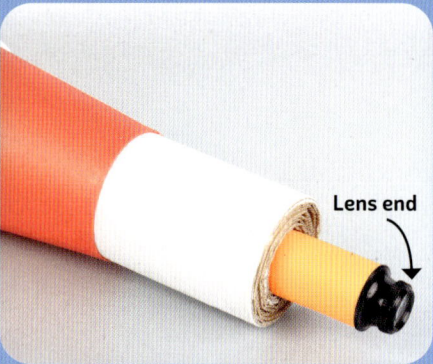

5 Slide the eyepiece tube into the hole, with the lens end on the outside. Make sure you can pull the tube in and out to focus.

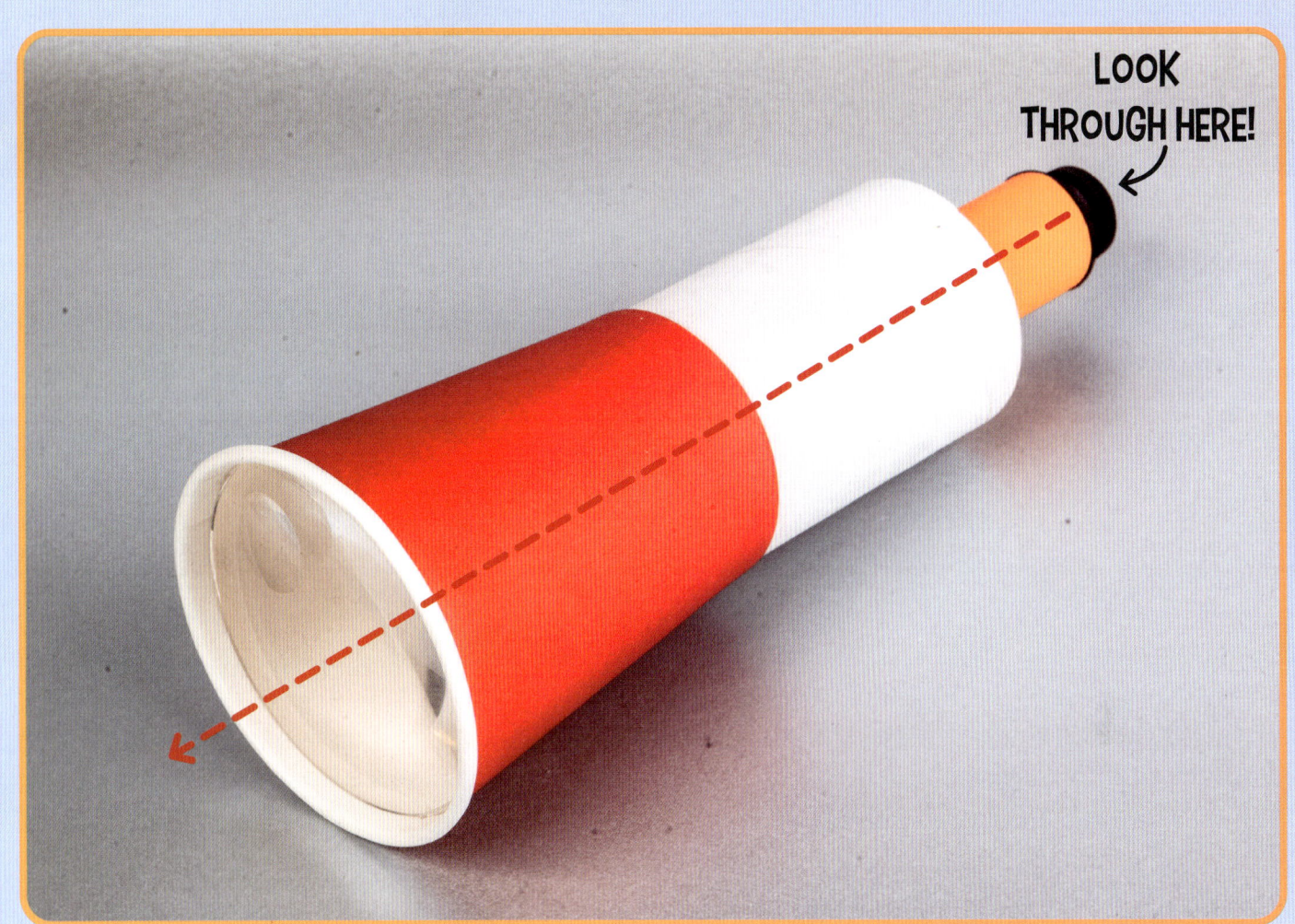

LOOK THROUGH HERE!

LOOK THROUGH THE EYEPIECE AND PULL THE TUBE IN AND OUT TO FOCUS THE TELESCOPE!

DID YOU KNOW?
Galileo was not the inventor of the telescope, but he was one of the first people to use one to study the sky. He was born in the 1500s and made many discoveries about distant planets, including the rings of Saturn and four of Jupiter's moons! It's thought he used a refracting telescope.

NOCTURNAL TIMEPIECE

This awesome **astronomical** instrument can be used to tell the time with the stars! Follow the "How to use" guide on the opposite page to do it yourself.

Use the QR code to access the template you need.

WHAT YOU NEED:
- Thin card
- Corrugated cardboard
- Felt-tip pens or pencils

TOOLS:
- Pair of scissors
- Strong craft glue
- Ruler

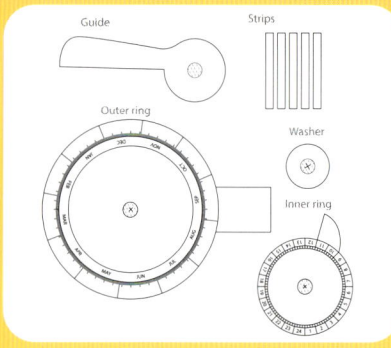

1 Print, copy, or trace the shapes from the template onto the specified materials and cut out.

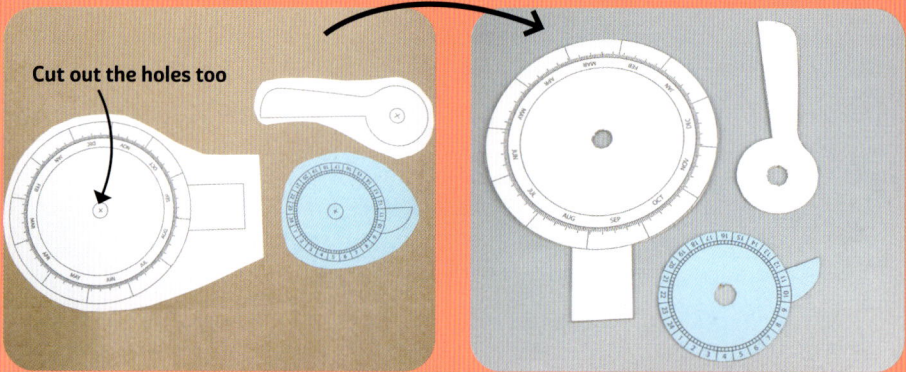

2 Glue the inner ring, outer ring, and guide template pieces onto thin corrugated cardboard and cut them out as shown. Don't glue the washer onto the corrugated cardboard; just set it aside until step 6.

3 Fill in both sides of the five strips of card with felt-tip pens or pencils, per the template instructions.

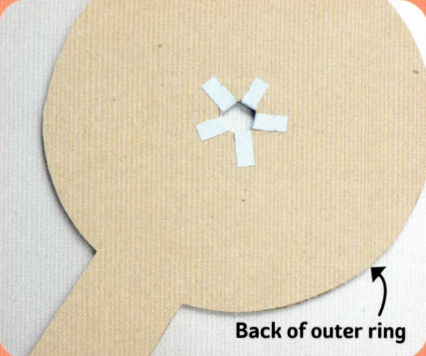

4 Thread the strips through the hole in the outer ring. Glue 12mm of each to the back as shown.

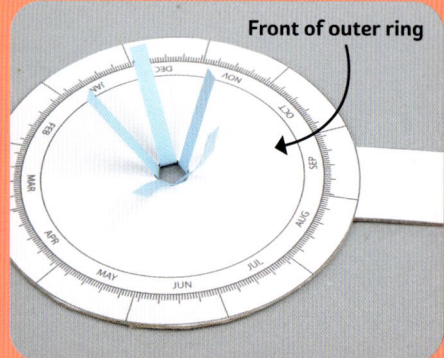

5 Flip the model over, straightening the long strips upward as shown.

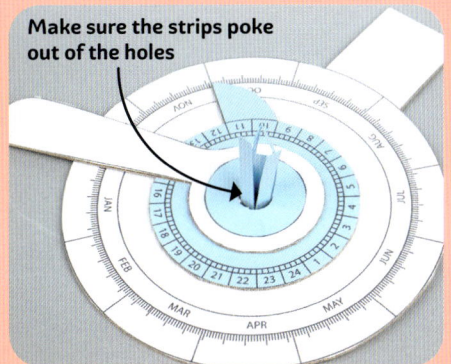

Make sure the strips poke out of the holes

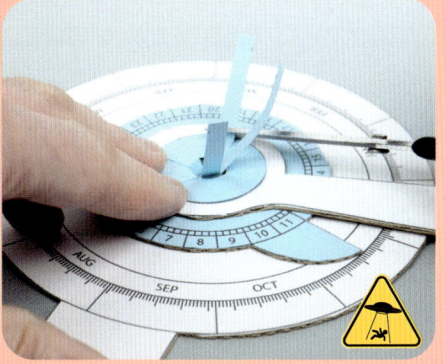

6 Thread the inner ring, guide, and washer on top of the outer ring in that order.

7 Fold the strips down and cut off any excess so they aren't longer than the washer. Glue them down.

TOP TIP

Once assembled, you should be able to twist each piece of the nocturnal independently.

How to use your nocturnal:

1. Turn the inner ring so that the finger is pointing toward the current date.

2. Hold the nocturnal upright and sight the North Star through the middle hole.

3. Rotate the guide to line up the flat edge with the two edge stars of Ursa Major. Read the time from the inner ring.

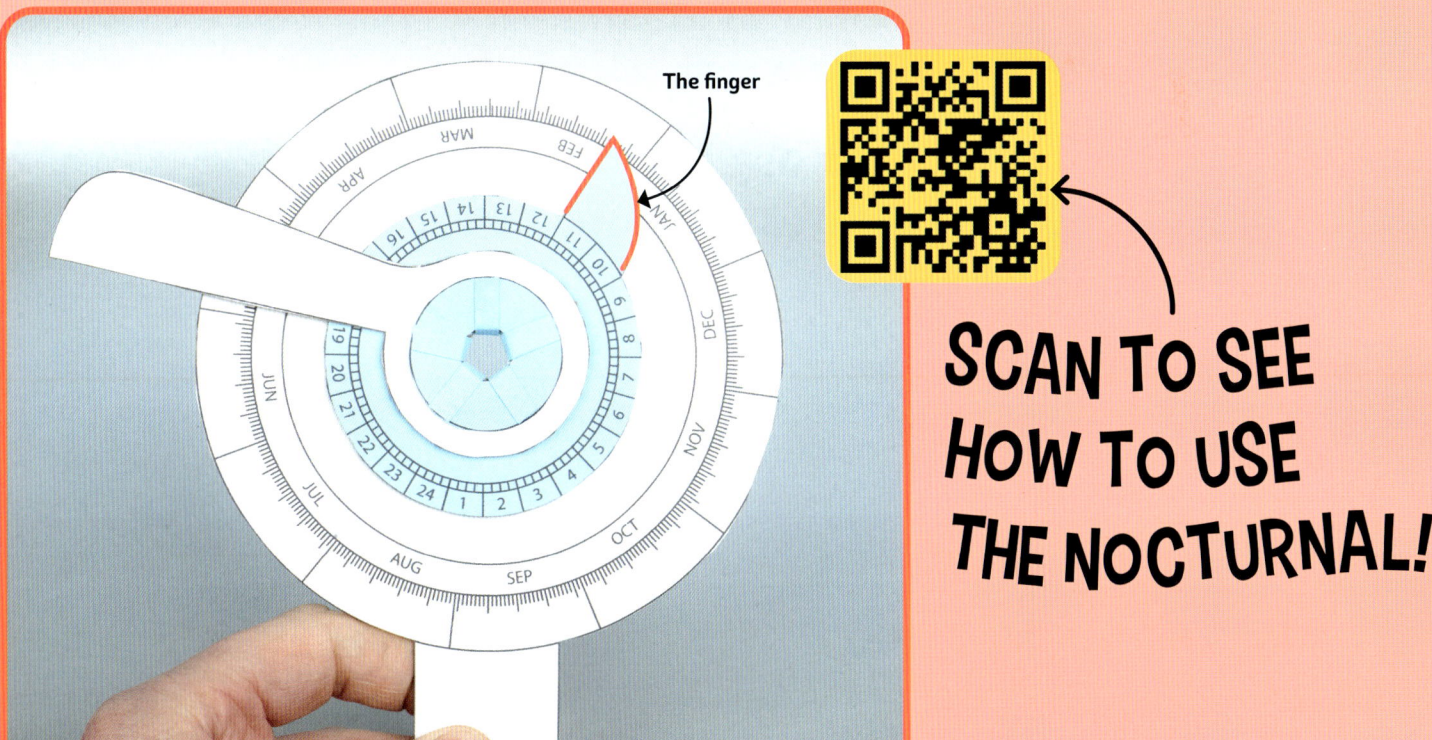

The finger

SCAN TO SEE HOW TO USE THE NOCTURNAL!

21

CONSTELLATIONS

Make these pocket-size **constellations** to help you learn and identify the beautiful stars in the sky.

Use the QR code to access the template you need.

WHAT YOU NEED:

- Poster tube 80mm in diameter
- Plain paper
- Black card
- Corrugated cardboard
- Torch or mini battery-operated lights

TOOLS:

- Pair of scissors
- Strong craft glue
- Embroidery needle or push pin

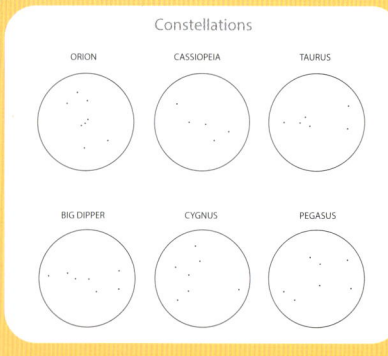

1 Print, copy, or trace the shapes from the template onto plain paper and cut out. Choose one constellation to start with.

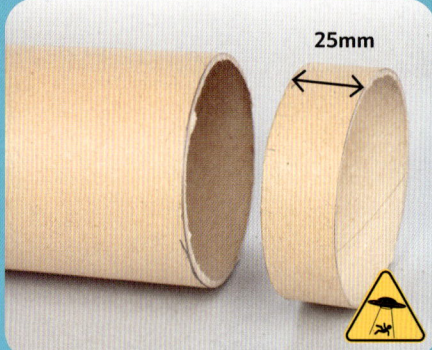

2 Ask an adult to cut a 25mm wide ring out of a poster tube (see page 7). **Don't have a poster tube? Use an empty tape roll.**

3 Hold the constellation over black card. Pierce the stars with a needle or push pin so the holes go through both layers.

4 Cut out the black card constellation circle.

5 Glue the constellation circle to the end of the poster tube ring and glue the name to the side of the tube. Why not decorate it, too!

6 When you're outside stargazing, shine your torch through the back of the tube to help you identify constellations.

Place some battery-operated lights inside and turn on to view the constellations whenever you like!

Try this craft with all the constellations provided!

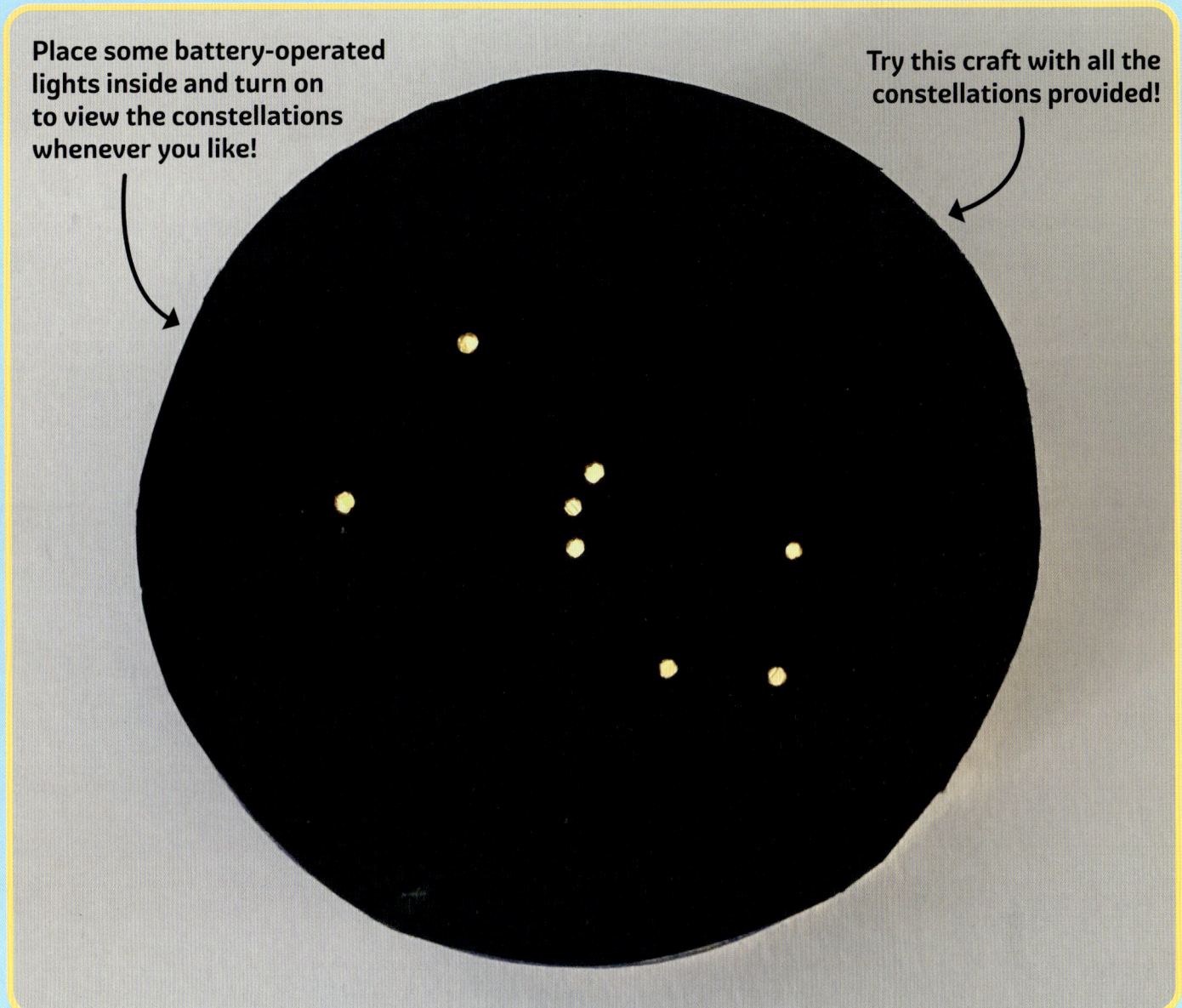

LOOK FOR THE CONSTELLATIONS ON A CLEAR AND STARRY NIGHT!

DID YOU KNOW?
If you looked at the sky every night, you would notice that the constellations change throughout the year. This is because the Earth is constantly moving around the Sun, whereas the faraway stars are not.

EQUATORIAL SUNDIAL

Sundials were used to tell the time long before clocks were invented. They use the Sun's shadow to show the time throughout the day.

Use the QR code to access the template you need.

WHAT YOU NEED:
- Poster tube 80mm in diameter
- Thin white card
- Bamboo skewer
- 2 pieces of corrugated cardboard

TOOLS:
- Pencil
- Ruler
- Strong craft glue
- Pair of scissors
- Semi-circular (180 degree) protractor

TOP TIP

If you don't have a poster tube, you could use an empty tape roll!

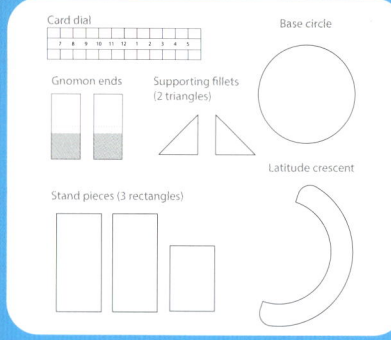

1 Print, copy, or trace the shapes from the template onto the specified materials and cut out.

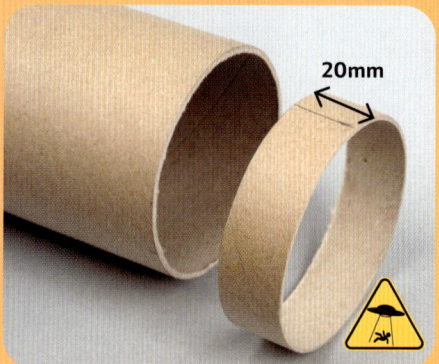

2 Ask an adult to cut a 20mm wide ring out of a poster tube (see page 7).

3 Cut the poster tube ring in half to make two semi-circles. Then, glue the two together at a 90-degree angle as shown.

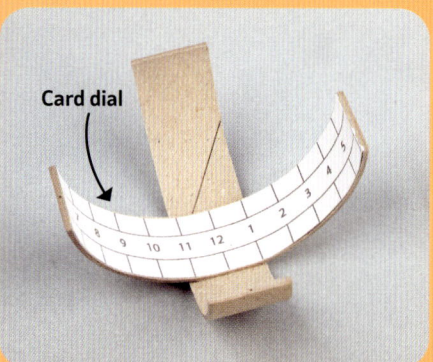

4 Glue the card dial to the top tube as shown, then set aside until step 7.

5 Take one of the gnomon end rectangles and glue two corners together as shown. Repeat for the second gnomon end.

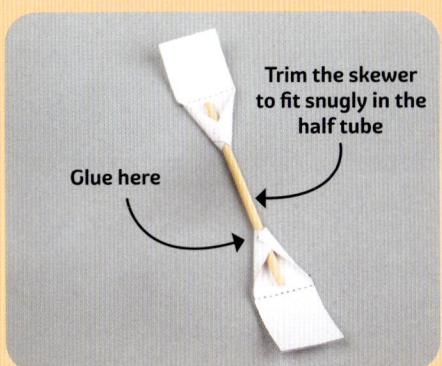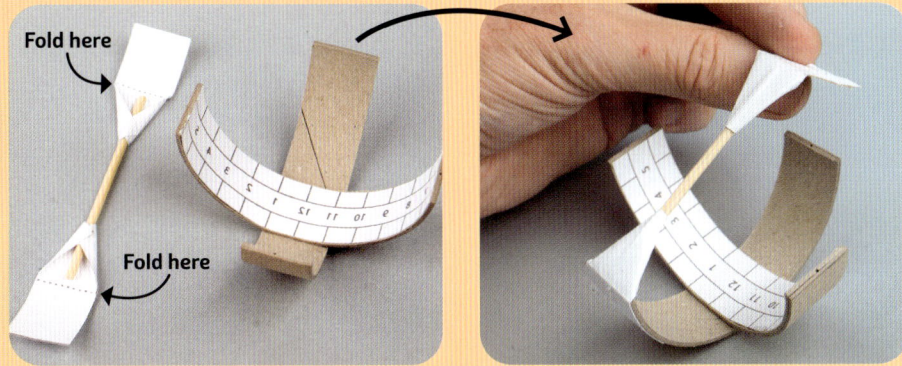

6 Once dry, thread a trimmed skewer into the gnomon ends as shown. Glue in place.

7 Fold inward along the dotted lines. Re-open and apply glue to the surface, then turn over the gnomon and attach each end to the tube without the dial. Set aside until step 11.

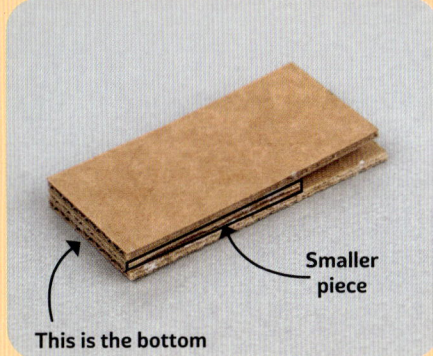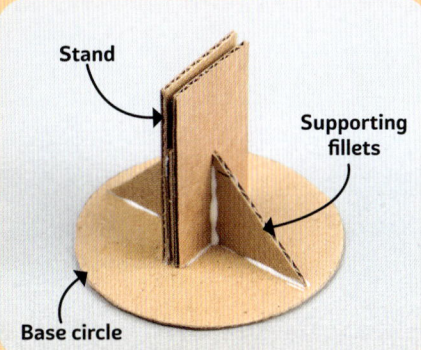

KEY WORD

LATITUDE

The line around the middle of the Earth is called the equator. Imagine a line wrapped around the Earth where you are. The angle between this line and the equator is your latitude. You can find your latitude with a quick search online.

8 Sandwich the small stand piece between the two bigger pieces. Line up the bottoms and glue together to make the stand.

9 Hold the stand upright in the middle of the base circle. Place the supporting fillets on each side of the stand and glue in place.

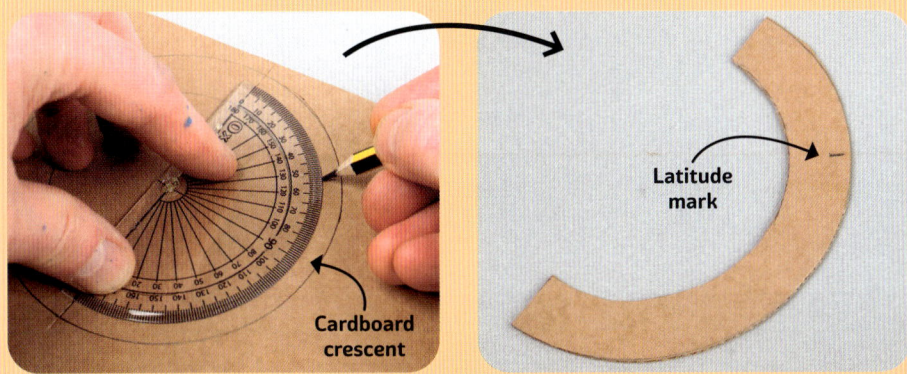

10 Find out the latitude of the place you live and mark that angle on the cardboard crescent from the template, with the aid of a protractor.

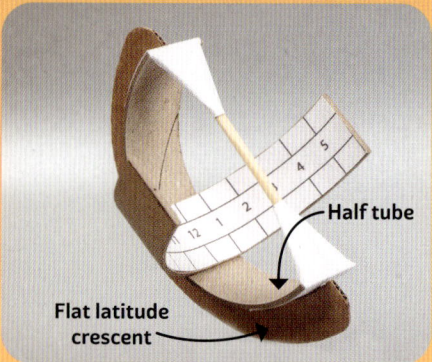

11 Glue the flat latitude crescent to the back of the back half tube as shown.

12 Once everything is dry, fit the sundial into the stand. Make sure that the latitude semi-circle is vertical and the latitude mark is fitted into the middle of the stand. Glue in place.

How to use your equatorial sundial:

1. Line up the sundial so that the gnomon faces north.

2. Read the time when the Sun shines by seeing where the shadow is cast on the dial.

3. If it's Daylight Saving Time, you may need to add or subtract an hour from your reading!

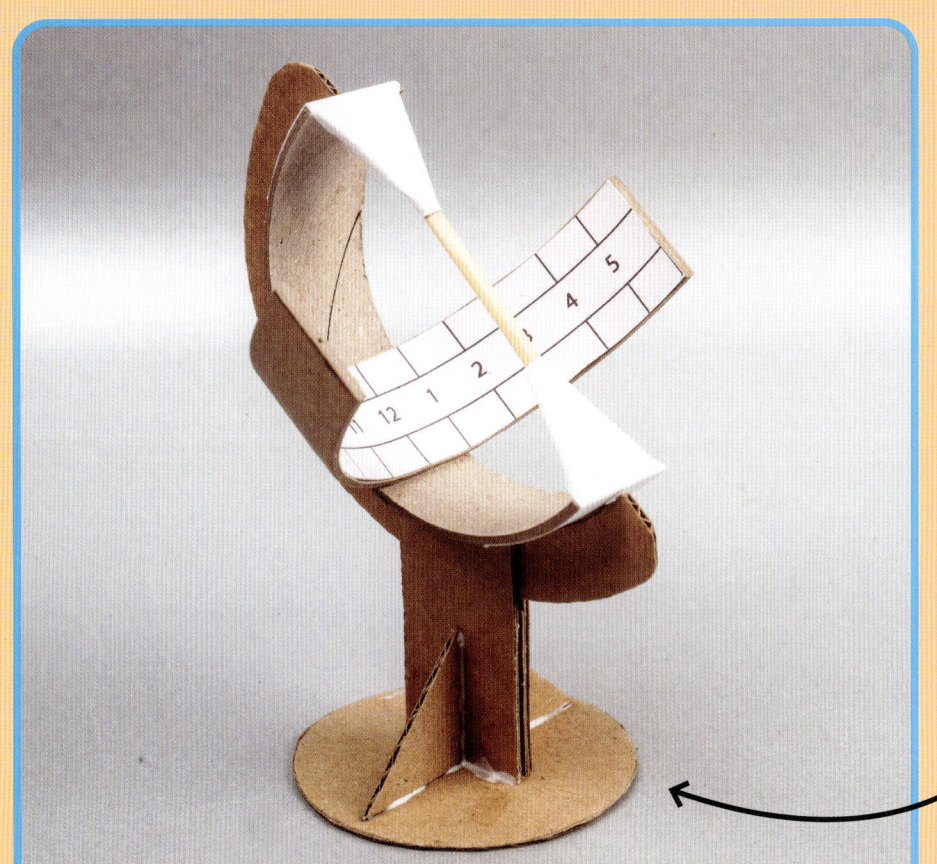

TOP TIP

Use a compass, or the North Star, Polaris, to find North.

USE YOUR SUNDIAL TO TELL THE TIME!

DID YOU KNOW?
The oldest sundial ever discovered dates to 1500 BCE. However, it's thought that as far back as 3500 BCE, people told the time by wedging a stick in the ground and watching the length of the shadow change. Sundials are much more decorative now, like this one here.

SUPER DISTANCES

Outer space is HUGE. The actual size of the universe is still a mystery, and a question that scientists may never be able to answer, but they have come up with clever ways of measuring the distance to the planets and galaxies that we have discovered.

MEASURING IN SPACE

Astronomical units (AU) are used to measure the distance between planets, 1 AU refers to the distance between the Sun and Earth, which astronomers have worked out is a massive 93,000,000 miles away from Earth.

Our solar system contains the objects which are closest to Earth, but even these are so far away that it's not possible to use Earth measurements to describe their distance. This is because they end up being huge numbers which are difficult to wrap our heads around!

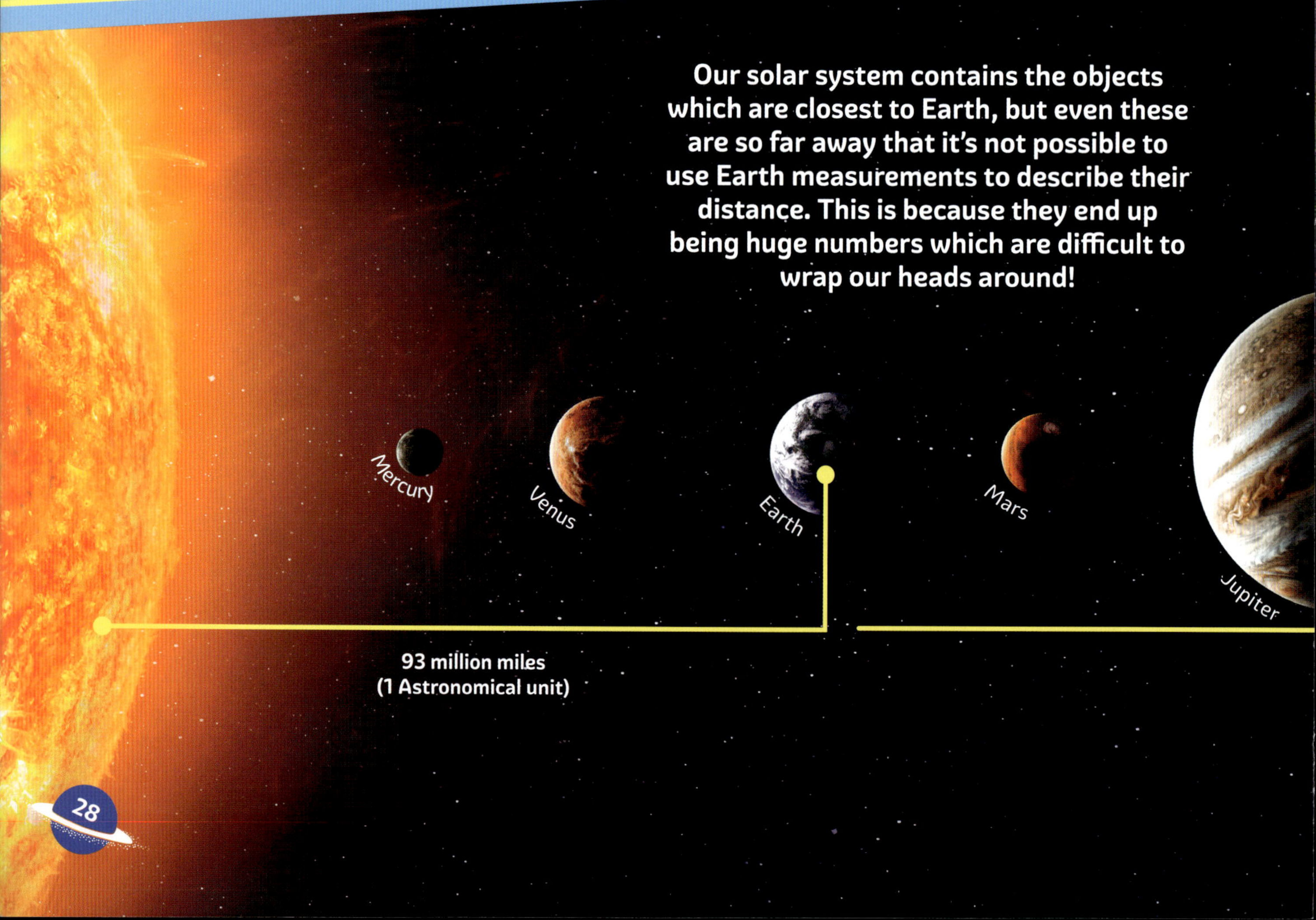

93 million miles
(1 Astronomical unit)

MEASURING IN LIGHT YEARS

Objects outside our solar system are even further away, so scientists use a different measurement called **light-years**. Light is the fastest travelling thing that we know of, so it's useful for understanding gigantic distances.

TELESCOPES

When we use powerful telescopes, like Hubble (page 30), to look at distant places, we're actually looking back in time! What we are seeing is what the star looked like years ago when the light started its journey!

Neptune is the planet in our solar system which is furthest from the Sun. It's an incredible 2.8 billion miles! This means that Neptune is 30 times further from the Sun than Earth is! Can you already see how using AU is easier than miles?

2.8 billion miles
(30 Astronomical units)

THE HUBBLE TELESCOPE

The Hubble Space Telescope is one of the most famous instruments used for studying space. It orbits Earth about 335 miles above the planet's surface. For over 30 years, it's beamed back incredible images of outer space.

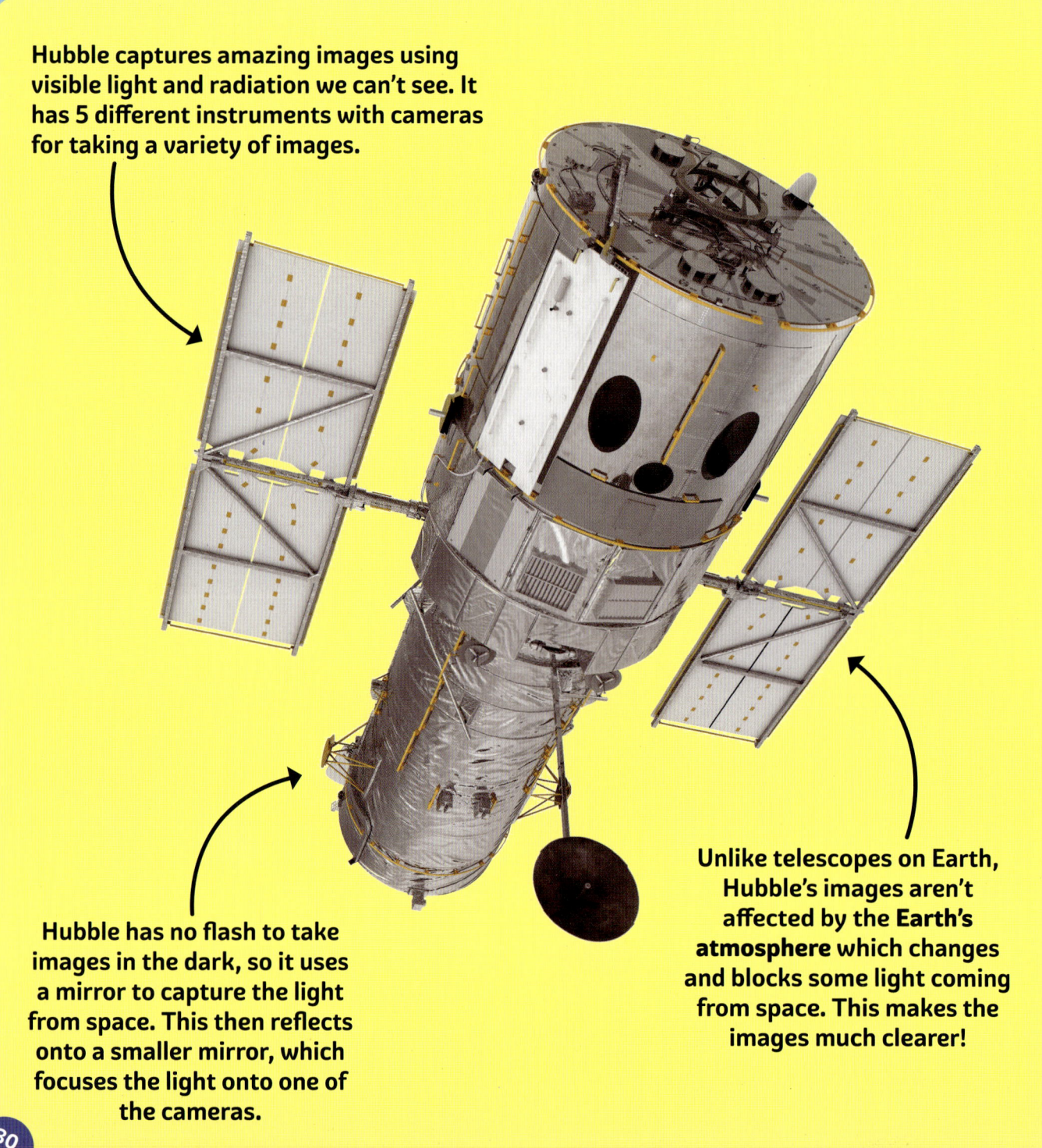

Hubble captures amazing images using visible light and radiation we can't see. It has 5 different instruments with cameras for taking a variety of images.

Hubble has no flash to take images in the dark, so it uses a mirror to capture the light from space. This then reflects onto a smaller mirror, which focuses the light onto one of the cameras.

Unlike telescopes on Earth, Hubble's images aren't affected by the **Earth's atmosphere** which changes and blocks some light coming from space. This makes the images much clearer!

SOUTHERN PINWHEEL
Below, is Hubble's view of a spiral **galaxy**, nicknamed the Southern Pinwheel, in the constellation Hydra. Being 15 million light-years away, it's very close and easy to see. The swirling blues and pinks are where stars are being born!

GLOSSARY

Astronomer – a scientist who studies outer space.

Astronomical - something which is to do with astronomy, the study of all objects outside Earth's atmosphere (see right).

Astronomical unit (AU) - a type of measurement that scientists use to measure big distances in space.

Constellations - groups of stars that form shapes and patterns in the night sky.

Earth's atmosphere – layers of gases surrounding Earth which are held in place by gravity (see below).

Galaxy – a huge collection of gas, dust, and billions of stars and their solar systems, all held together by gravity (see below).

Gravity - a pulling force that works across space. Objects don't have to touch each other for gravity to affect them. For example, the Sun, which is millions of miles away, pulls on Earth and the other planets and objects in the solar system to keep them in orbit.

Light-year - a unit of distance that shows how far light travels in one year.

Moon phases - the different shapes on the Moon we see in the sky. These follow the same cycle every month.

Orbit – the repeated path taken by one object circling around another object in space.

INDEX

C
constellations 22-23, 32

E
Earth 9, 25, 28-29, 30-31, 32

G
Galileo 19

H
Hubble telescope 29, 30-31

I
inclinometer 10-11

L
landers 38-39, 46-47
latitude 29, 30

M
Moon 9, 12-13, 14-15, 17, 19
Moon phases 12-13, 14, 32

N
nocturnal 20-21

P
pendulum 10-11
planets 9, 15, 17, 19, 28-29, 30

S
solar system 9, 28-29
stars 5, 9, 10-11, 15, 20-21, 22-23, 31
Sun 8-9, 17, 23, 24, 26-27, 28-29, 32
sundial 24-15, 26-27

T
telescope 9, 15, 16-17, 18-19, 29, 30

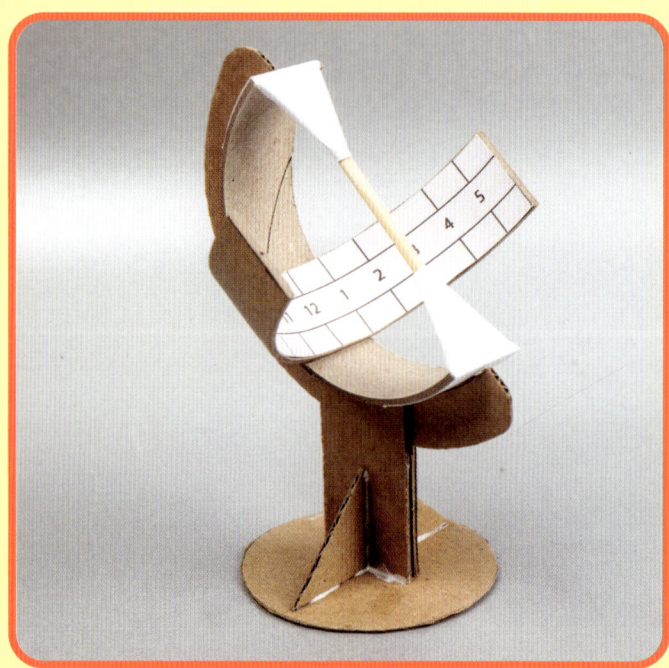

PICTURE CREDITS:

(Abbreviations: t=top, b=bottom, m=middle, l=left, r=right, bg=background)
Shutterstock: 3DMI 30m; Alex Terentii 28-29bg; FishCoolish Astronaut character throughout; Franco Tognarini 19b; Klyaksun spaceship/rocket throughout; Lukasz Pawel Szczepanski 26-27b; NASA Images: 8b, 31b; Nunik5 26br; Paulista 29tr; Solarseven 29tl; Vector Tradition 28tr; Vovan 17tr.

Every effort has been made to trace the copyright holders, and we apologise in advance for any unintentional omissions. We would be pleased to insert the appropriate acknowledgements in any subsequent edition of this publication.